THE SECRET BEHIND THE UNIVERSE

(Cosmology in Vedanta – The Physics Correlation)

Second Edition

C Radhakrishnan

Gopal K R

Published by Hi-Tech Books

Azad Road, Kaloor, Kochi 682017, India

The Secret Behind the Universe (Cosmology in Vedata – The Physics Correlation), a work comprising of an article in physics and vedantic cosmology, written in English by C. Radhakrishnan and Gopal K.R.

First published in July 2016, Second edition

Cover picture: An interacting colossus, taken by the NASA/ESA Hubble Space Telescope's Wide Field Planetary Camera 2 (WFPC2), shows a galaxy known as NGC 6872 in the constellation of Pavo (The Peacock). Image credit: ESA/Hubble & NASA. Acknowledgement: Judy Schmidt (geckzilla.com)

ISBN-13:978-1535342513

ISBN-10:153534251X

About the authors

C Radhakrishnan is an established author in the Malayalam language. A scientist-turned writer, he has contributed to all branches of creative literature besides popular science. He has been honoured by India's National Akademi of Letters, the Kerala Sahitya Akademi and almost every other body promoting creative literature of the language he writes. He is the winner of Ezhuthachan Award (the highest recognition given to authors in the Malayalam language), and also the Murtidevi Award of the Bharatiya Gnanpith Trust for the best creative work in Indian languages. A multi-faceted personality, he has worked as scientist, popular science journalist, film director, columnist, and editor of national and literary journals.

Born on 15 Feb 1939 in Chamravattom village in Kerala, C. Radhakrishnan became the Winner of Best Student Gold Medal at Achutha Varier High School, Ponnani, winner of Madras University Sharpe Scholarship - Inter: Zamorin's College, Calicut, and Winner of Top Rank Gold medal - Graduation: Physics, Zamorin's College, Calicut - I. PG: Post Graduation - Applied Physics; Research (Astrophysics).

The author also won in every literary competition during his academic career. He translated Daniel Defoe (M. Flanders) and Lincoln Barnett (The Universe and Doctor Einstein) at age seventeen, and David O.Woodbury's Outward Bound for Space in 1962.

C. Radhakrishnan became scientific Assistant at Astrophysical Observatory, Kodaikanal in 1960 and became officer-in-Charge, World-Wide Seismology Centre, Pune in 1962. He helped launch and establish Science Today (Times of India), the first popular science magazine of India, at a young age of 24, and later held senior positions with national print media organs. He was also the science Editor of Link Magazine and Patriot Daily, Delhi, 1968-'72.

His research on problems fundamental to astrophysics gave rise to a monograph titled "Unity of Space-Matter Manifestations", published as early as in January 1988. "Stuff and Style of the Universe", a book which is a popular science elaboration of the monograph, was published Nov. 2002 by Hi-Tech Books. The first Scientist-President of India, Dr. A. P. J. Abdul Kalam, expressed appreciation of the work done, and the model attracted world attention. An article based on the book was also published in 'Science India' (ISSN 0972-8287) in two issues, which came out on September 2012 and October 2012 (Vol. 15 No. 9 and Vol. 15 No. 10 respectively). Continued researches resulted in this book "The Secret Behind the Universe", published July 2016, in which the original ideas were further refined and updated. This was the work of a father and son team, as the author's son Dr. K. R. Gopal also contributed in research and writing.

A condensed version of the physics section of this book was published in January 2017 as an article titled "Avyakta: The Fabric of Space" in the Prespacetime Journal Vol 7 Issue 16, and it was written jointly by the author C. Radhakrishnan and his son Dr. Gopal K. R. (Prespacetime Journal (ISSN: 2153-8301), QuantumDream Inc., P.O. Box 267, Stony Brook, NY 11790-0267, USA; is a physics journal which focuses on the origin, nature and mechanism of spacetime and its possible connection to a prespacetime; and models and experimental results on elemental particles, fundamental forces including gravity and related topics.) The entire article is available online to read or download from the journal website, and can be accessed through the author's site: http:://www.c-radhakrishnan.info.

CONTENTS | Pages

PART 1

Physics

THE NEXT STEP

Abstract

A new concept, the 'Fabric of Space' is introduced to complement the Standard Model and explain gravitation. It is attempted to be shown that the fabric of space provides substrate for all the fields including electromagnetic field, gravitational field and Higgs field, without conflicting with current concepts. Dark matter and virtual particles gain new meaning with the idea of the fabric of space, and cosmic inflation can be seen in fresh light. Convincing explanation is provided for matter/antimatter asymmetry and generations of matter. The deep concept is also attempted to be presented in the simplest of terms possible.

Introduction to Avyakta

It is assumed that a fabric of special nature pervades the entire universe as a continuous web. It is everywhere. We can introduce a term for it - 'Avyakta'.

One may wonder where this 'fabric' is. *Avyakta* cannot be seen with the naked eye. Why don't we feel it? Is it separate from the sun and all the planets? The answer is that *avyakta* is not separate from any 'thing' in it, because all types of materials we see are its own manifestations. And space only means a place which is devoid of matter. *Avyakta* is everywhere.

This background fabric is assumed to have a unique characteristic - it is pliable in its own way so that it can spread or shrink in a general fashion. The behavior can be visualized as similar to an elastic material – when it is stretched, it has a tendency to return back to its original state. Also when it is compressed, its tendency is to return to its flat state.

Let this fabric of space (avyakta) be called S for short. Before venturing any further, fresh terms are needed. Space is what we observe if we look around. S is the fabric that goes into its making. In fact, no one ever sees S. Expansion and contraction are terms that normally lead and connect one to 'known' physical parameters like pressure, density, elasticity and so on. The characteristics of S being basically different, they call for an entirely new genre of terms. Therefore when the fabric of space is compressed, we can call it Tough, and it has a tendency to spread. Similarly when it is stretched, let us call it humble, and its tendency is to shrink. Flat means it is in the ground state. At any time at any given point, S is either Tough or Humble or Flat. If Flat, it will not expand or shrink, it will remain idle.

Toughness and humbleness indicate action-potential of mutually opposite nature. The word 'Vigour' can be used to indicate this potential.

Avyakta is tough at this stage of the universe, but it is declining in toughness. The decline started from the Big Bang, and the background fabric has consistently spread out but has not yet reached the Flat state.

Let us consider a particle as a small volume of S with an additional toughness, surrounded by S of normal background toughness. It is assumed that at the subatomic (quantum) level, this extra toughness which is the particle exists as a standing wave of the fabric.

These assumptions correlate and also provide explanation as to why every particle in the universe has wave behavior. Matter waves are discussed next.

SECTION 1

Fundamental Particles, Forces and Fields

1.1 Matter Waves

According to the de Broglie hypothesis, every object in the universe is a wave. Matter waves are a central part of the theory of quantum mechanics. The position of any particle is described by a wave function. It is already known that a beam of electrons can be diffracted just like a beam of light. (Diffraction is the bending of light around the corners of an obstacle or aperture into the region of geometrical shadow of the obstacle, and interference is a phenomenon in which two waves superpose to form a resultant wave of greater, lower, or the same amplitude.) Wave behavior of matter was first experimentally demonstrated in electrons, by the Davisson–Germer experiment. *Subsequently wave behavior was also confirmed for other elementary particles and even atoms.* Experiments with Fresnel diffraction and atomic mirror on neutral atoms confirmed the existence of atomic waves which undergo diffraction and interference. The wave behavior of matter is also crucial to the modern theory of atomic structure and particle physics.

In the sun, hydrogen is converted to helium via proton-proton chain reaction, but the temperature of its core is a little low to overcome the repulsive Coulomb force for two protons to come together. It is through the phenomenon of tunneling that the protons overcome this barrier. (Tunneling refers to the quantum mechanical phenomenon where a particle tunnels through a barrier that it cannot surmount if it was only a solid particle.) Quantum mechanics explains that the wave function of a particle summarizes everything that can be known about a physical system. Quantum tunneling correlates with a dynamic wave model and explains a lot many properties of atomic particles.

1.2 Movement through avyakta

What makes a particle of any kind shift position? Attraction by another particle is one reason, the opposite of it, another. Body impact is yet another.

A great deal of effort has been made to better understand the rules of this game so that the movements of the players can be explained in stricter terms. Why not study the terrain of the 'playground' also? Let us see what role, if any, S plays in the act of 'bodies' moving in it.

In avyakta, any particle is an integral part of its surroundings. When a particle 'moves', the S involved in its oscillation does not. Only the energy and the consequent oscillating act shifts to the adjacent S. The matter wave only develops an additional direction or vector component in its wave-form. This additional vector component is the reason why any matter wave, once it begins shifting, continues in that direction even though the force that made it move in the first instance has disappeared. The extra toughness or extra humbleness which is the physical content of the particle is assumed by the adjacent S, while the S which hosted the particle reverts to the surrounding background toughness. Therefore the 'content' in the particle is different after it moves. This is the mode of movement of waves in any medium – take for example the waves in water. Even though the wave is conducted from one place to another, individual molecules are not carried away with it, they just oscillate to conduct the energy of the wave.

Any 'persuasion' is, naturally, in the form of a force. When a particle is subjected to a force, there are just two ways in which the force can act on the particle - either directly to the body or through a 'field' in S. The force, if applied bodily, acts at a point on it. This in effect becomes an obvious attempt to deform it. The wave form that comprises the particle cannot help but react.

What decides the critical value of the force needed to move a particle? There are only two factors to be taken into account - the 'bulk' of the particle and the 'smoothness' of the terrain i.e., surrounding avyakta. The 'bulk' in this context means the toughness that has gone into the making of the particle. If the nature of the terrain remains the same, this alone decides the minimum of 'insinuation' needed to move the particle. The more the content, the greater the force needed to overcome the inertia, in direct proportion. But this is only as long as the terrain remains the same. In avyakta of a different toughness, the critical dose of force needed to shift the same particle will be different.

If the particle shifts once, it goes on shifting because its wave form has picked up an additional vector while getting rid of the deformity imposed on it by the force applied. Movement is kept on till it is undone by encounter with an equal and opposite force.

As the wave action alone is what shifts when a particle moves from one place to another, the content of S in the particle after transfer is entirely different. Just the action part of it gets transplanted. The event also can be viewed as a different location of S claiming the action. What takes place may be called an instance of transmigration of the wave form.

1.3 Gravitational field

The quark is the most elementary particle which participates in all four fundamental interactions, viz., electromagnetism, gravitation, strong interaction and weak interaction. It is the building block of all matter in the universe. Therefore we can investigate the quark to see how it is related to the gravitational field.

It is widely accepted that the internal wave structure of subatomic particles is complicated such that it cannot be explained in terms of any simple physical model. However, here we assume that the wave form of the T of the quark is spherical and oscillatory in nature, having an outgoing spiral component and an incoming spiral component, with a definite chirality. (A spiral oscillation with chirality is not difficult to visualize; for example the common sea shell has such a pattern.)

(An object or a system is *chiral* if it is distinguishable from its mirror image; that is, it cannot be superposed onto it, like left hand on the right.)

Inferences: The continuity of the particle with the surroundings is not broken any time because avyakta is a continuous fabric and the particle is not separate from it. Therefore during the phases of wave motion, corresponding to the incoming spiral oscillation phase, the decline in toughness in the fabric just outside, or the jerking-in, spreads to the surroundings of the particle. Similarly during outgoing spiral oscillation, outside the particle the gradient caused by the spreading out would be refilled. Both these oscillations establish themselves as minute concentric waves in the fabric around the particle.

Avyakta has been decreasing in toughness from the big bang onwards, but is still relatively tougher than that in the Flat state. As the surrounding fabric is tough, its readiness to spread is a lot more than its willingness to shrink. Therefore even though the oscillations in the particle are resonant, the surrounding avyakta always favours 'filling in' or the incoming oscillation of the particle. This effect can be taken as almost negligible in the majority of oscillations pertaining to a tiny sub atomic particle (the quark), as its internal wave motion is well balanced within the baryon. But in a larger conglomeration the situation is different. The tendency of the surrounding avyakta to favour incoming oscillations reinforces through all the combined oscillations of its constituent particles as a set of concentric waves around the assembly.

Therefore around a large assembly of particles, there is an in-jerking wave in avyakta due to decrease in toughness that spreads forcefully, followed by the reluctant and slower making up of it. When there is another particle in its vicinity, the sudden decline in toughness acts on it, tending to pull and deform the wave form of the guest particle in that direction. This amounts to an additional vector in the internal oscillation of the guest particle in the direction of the deformation. As a result the particle is attracted towards the host.

Isn't there a lot more to this mystery? The waves that spread affect the host particle and the surrounding particles in several ways at once.

1. Gravity is transmitted across by the gravitational waves with a speed not more than what any wave can move through avyakta. Therefore gravitation cannot act instantaneously at a distance.

2. The spiral during the time of the incoming component in the host particle also gives a very weak spiral component to the gravitational wave. It tends to rotate the guest in the direction of the incoming spiral of the ripple. Every part of the body subjected to the pull of the curvature of the fabric also experiences the tangential component of the spiral. Therefore, a free-falling body tends to circle the attracting body instead of coming straight to it. The weak spiral

component also causes the host body to rotate on itself (if its mass is equally distributed / other things remaining the same). The arms of the spiral that reach its front are stronger than the arms that reach its back, and reach thereabouts 'later'. Hence there is a torque.

3. As and when particles are 'sufficiently' pulled in, they eventually conglomerate. If the particles in a conglomerate are free to move within it, gravitational pull induces every particle to place itself as close as possible to every other of the lot. The assembly therefore assumes spherical shape.

4.When there are many bodies orbiting a massive attracting body, gravitational force between the 'free-fallers' tends to align them in a common plane; as that is how every one of them can get closest to everyone else in the course of their orbital motions.

5. Though the waves ideally reaches infinite distances, in any particular case there arises a definite limit beyond which it falls short of strength to overcome the 'inertia' of even the smallest of particles.

6. When there is a large spherical assembly of particles, the gravitational waves around them reinforce each other most at the centre of the sphere. If the body is large enough, the particles so far fortified by it begin to get vulnerable. If the size of the conglomeration crosses a certain limit, the background avyakta itself begins to go into contraction phase initiating gravitational collapse.

7. Gravitational field is not manifest for instrumentation or experiments as it is the sole property of the background avyakta. Due to the weak nature of the force and the invisible background, physics has major difficulties in finding the representative boson for it (graviton).

(In this article the fourth dimensional aspect for avyakta is not explored, though mathematically any number of dimensions is possible.)

1.4 Mass

Mass is a measure of an object's resistance to changing its state of motion when a force is applied. It is determined by the strength of its gravitational attraction to other bodies, its resistance to being accelerated by a force, and the energy content of a system. For subatomic particles, the energy content is a very important contributor. Mass is proportional to the amount of energy contained within the body, using $E = mc^2$.

The extra toughness which is the content of the particle with its independent wave motion provides it a place in the fabric of space. Movement in any direction would mean addition of further vector component to the wave form and therefore require application of force. Its incoming oscillation makes it involved in gravitation. Those which have an additional, independent and stable content of tough or humble S involved in its wave-action will have mass. This applies to however tiny they are; e.g., neutrinos.

The dynamic involvement with the surrounding avyakta indicates problems related to fixing the exact mass of subatomic particles because some of the wave components are manifested only during certain instances of time during their oscillation, and some parts may remain always un-manifested. The wave gradually goes into the flat stages during its oscillation from humble to tough; for an observer, the wave would seem to 'fade' into and 'come back' from avyakta. Also there is possibility of add ons as the essential content is the same in the particle and the surroundings. This is demonstrated in the phenomenon of neutrino oscillation; there are actually three types of neutrinos based on their mass and interaction – the tau, muon and electron neutrinos. Yet they oscillate between the different types or inter-change themselves.

It can also be inferred that mass does not relate to the 'amount' of S in the particle. The mass of the tough particle is the same as that of the humble particle (antiparticle), because in both extremes of tough and humble S, the vigour or its tendency to react is the same. Therefore energy is the same in both extremes irrespective of the 'content' in the oscillation. Apparent mass would be the same in both particle and antiparticle. (Antiparticles are discussed in a later section.

1.5 The fields in classical physics

Magnets have an area around them in 'empty' space where a force exists which is strong enough to either attract or repel another magnetic substance. Magnetic fields in 'space' even give power to other particles that come into contact with the field. Magnetic field lines show the direction of the force and its strength.

Magnetism was once thought to consist of currents of aether. Even after the Aether theory was proved wrong by the Michelson-Morley experiment, the vector algebra still described electromagnetic phenomena adequately so that the disappearance of the aether left the concept of magnetic flux much unchanged. It was forgotten that 'flux' meant flow.

In the modern framework of the quantum theory of fields, as a field contains energy it is assumed that its presence eliminates the vacuum. "A particle makes a field, and a field acts on another particle, and the field has such familiar properties as energy content and momentum, just as particles can have". And a field particle is assumed to represent the field, namely a boson. The question of the flux is got around by assuming that the electromagnetic field consists of 'virtual' photons. It had long ago been accepted that the electromagnetic wave need not have a medium for propagation. Therefore the background supporting all this drama could be forgotten. But there is one problem. What if that 'vacuum' behind the field itself proves that it is not empty?

1.6 The 'vacuum' state – the 'zero' point energy of all fields in space

In quantum field theory, the vacuum state is the quantum state with the lowest possible energy where the fields are quantized. So this vacuum state is absolutely empty and should have zero energy; but the actual situation is found to be different. ***Even in the vacuum state there is energy which has measurable effects*** – quantum vacuum zero-point energy, the lowest possible energy that a quantum mechanical physical system can have; the energy of its ground state. ***It is the zero-point energy of all the fields in space, which in the Standard Model includes the electromagnetic field and the Higgs field.*** In fact, this energy of a cubic centimetre of 'empty' space has been calculated figuratively to be one trillionth of an erg.

And there is even more to it. In quantum field theory, the 'empty' fabric of space is visualized as consisting of fields, with the field at every point in space and time being a *quantum harmonic oscillator*, and not only that; the neighbouring oscillators are also always in a constant state of interaction. Therefore from every point in 'empty' space, there is a contribution of energy.

It gets even more interesting with fluctuation of this energy in vacuum. A quantum vacuum fluctuation is the temporary change in the amount of energy in a point in space. That means that in Quantum Electro Dynamics, conservation of energy can appear to be violated.

These fluctuations are taken by QED to allow for the creation of 'virtual particles', in an effort to explain vacuum polarization. (Vacuum polarization describes a process in which a background electromagnetic field somewhat magically produces virtual electron–positron pairs that change the distribution of charges and currents that generated the original electromagnetic field. It is also sometimes referred to as the "self energy" of the photon.) The term 'vacuum fluctuations' refers to the variance of the field strength in the minimal energy state, and is described picturesquely as evidence of "virtual particles".

According to present-day understanding of what is called the vacuum state or the quantum vacuum, it is "by no means a simple empty space", and again: "it is a mistake to think of any physical vacuum as some absolutely empty void."- (*Christopher Ray (1991). Time, space and philosophy. London/New York: Routledge. Chapter 10, p. 205. ISBN 0-415-03221-0.*) "According to quantum mechanics, the vacuum state is not truly empty but instead contains fleeting electromagnetic waves and particles that pop into and out of existence."- (*Astrid Lambrecht (Hartmut Figger, Dieter Meschede, Claus Zimmermann Eds.) (2002). Observing mechanical dissipation in the quantum vacuum: an experimental challenge; in Laser physics at the limits.*)

As per the explanation in QED, these "virtual" particle–antiparticle pairs (leptons, quarks or gluons) created out of a 'vacuum' conveniently annihilate each other after the required time. They can carry various kinds of charges, such as colour charge. Charged virtual pairs such as electron–positron pairs can act as electric dipoles, and in the presence of the electromagnetic field around an electron they reposition themselves, thus partially counteracting the field. All these explanations are to provide reason as to why the field is weaker than would be expected if the vacuum were completely empty. The reorientation of the short-lived particle-antiparticle pairs is taken as the cause for vacuum polarization.

Explanations by the concept of avyakta: Without realizing the dynamic involvement of avyakta in these effects, the only feasible explanation is by attributing everything to virtual particles, appearing by pure magic out of nowhere and excusing themselves on account of Heisenberg's uncertainty principle.

Avyakta is tough at this stage of the universe, therefore its tendency is to spread out and fill oscillations or support wave forms already existing in it. This is the dynamic background which contains all the virtual particles, like a sea of background waves interacting with already formed ones. The same has the tendency to alleviate the inadequacy and dampen the field.

The magnetic field is magnetic resonance in the surrounding S, produced as the unpaired electron precesses in the atom. Energy, or flux, or virtual photons in QED, flows from one pole to the other through the fabric in the form of resonance waves. A cross section of the electron along with the field is represented by the flux in the field arranged in concentric rings with alternating areas of increasing and reducing strength. The ring is resonance due to unpaired precession of the electron, and the alternating waves are due to its unpaired oscillation phases.

The electric and magnetic fields are similar to the gravitational field. The oscillating toughness of the baryon correlates with gravitational waves; similarly the structure of the electron correlates with the magnetic field.

As the magnetic and electric fields are supported by and formed from the background avyakta, this indicates problems to assigning values to the particles because of their continuous interaction with the background; at the subatomic level everything would cater to 'laws' only within limits. The g-factor would never show a normal result if only the original model of the electron particle is considered.

The anomalous magnetic moment of the electron, and indeed all composite particles, are attributed in QED to the effect of vacuum energy. (The magnetic moment, also called magnetic dipole moment, is a measure of the strength of a magnetic source. For particles such as the electron, the expected result differs from the observed value by a small fraction of a percent. The difference is the anomalous magnetic moment.) The differences in energy between the 'equal' energy levels of the hydrogen atom (the Lamb Shift) is also because of fluctuation in the position of the electron due to interaction with the vacuum energy fluctuation, by explaining that there exist small zero-point oscillations that cause the electron to execute rapid oscillatory motions; again pointing out to the nature of the background. The electron is 'smeared out' and the radius is changed.

1.7 Virtual particles

The present established concept of ignoring the dynamic background has compelled physics to explain a lot of phenomena based on 'virtual particles', so that their list now seems endless. The things the virtual particles can do also appears to be 'super-particulate' in many respects because, without the concept of the dynamic interactions with the fabric of space, any particle that is actually observed never precisely satisfies the conditions theoretically imposed on regular particles.

The advantage of assuming virtual particles is that they can be presented in convenient combinations that mutually more or less nearly cancel so that no actual violation of the laws of physics occurs in completed processes. They are also visualized as 'conceptual devices' that do not even need the same mass as the corresponding real particle. If the mathematical terms that are interpreted due to the presence virtual particles are omitted from the calculations, an accurate result is not obtained in many situations. This is because there is presently no concept of the dynamic background. There are 'virtual particles' for explanatory visualizations in almost everything in quantum mechanics.

Examples of explanations with virtual particle help include the Coulomb force due the exchange of virtual photons, the magnetic field caused by the exchange of virtual photons, electromagnetic induction, Casimir effect (where the background field causes attraction between a pair of electrically neutral metal plates, and the Van der Waals force, which is partly due to the Casimir effect between two atoms), vacuum polarization and Lamb shift, the spontaneous emission of a photon from an excited atom or excited nucleus in which the background of QED 'vacuum' mix with the excited state of the atom to release the photon (Spontaneous emission in free space depends upon vacuum fluctuations to get started.), Hawking radiation from black holes etc.

The weak nuclear force mediated by virtual W and Z bosons is another example. In β- decay, the neutron is converted to a proton plus an electron and an antineutrino. (We know that the neutron is composed of three quarks – two down quarks and one up quark. One of the down quarks converts to an up quark with the ejection of the electron. The resultant combination of two up quarks and one down quark constitute the proton.) In the movement of β decay, the background oscillation which is formed is complex in itself to permit the conversion of the free quantum from a quark within the nucleus in a way that makes it possible to adjust to the area around the baryon. The background slot is only for the briefest period of time as the force carrier for the ejection, conversion and stabilization of the electron. This is the W boson (W-) and the characteristics of the slot have already been derived by the Standard Model. This boson, whose job is to convert the tiny quantum from the quark to an electron, is a hundred times larger than the proton, why, even heavier than atoms of iron! It is irrational physics unless there is a dynamic background. The W boson is from the background avyakta; it is a virtual particle in that it is never 'independent' from the background, it has no 'existence' before and after the event and the whole 'set' converts as soon as it is formed. Experimentally the existence of the W boson is proved but it is never manifest.

The same is the case with Z boson. It is another heavyweight, but its job lies in elastic scattering of the tiny neutrinos. It is never manifest any time else and has no independent existence. Both the manifestation of Z boson and neutrino oscillation point to interactions with the background avyakta. The W and Z bosons are temporary condensations of the background fabric in the pattern conductive to the respective events.

The strong nuclear force: The force works on two levels – in level one it is the force that makes the gluons hold the three quarks together to make a proton or neutron; in level two it is the force that holds together the protons and neutrons to form the nucleus.

The gluons are assumed to be 'mass-less' particles which hold the quarks together. The strong nuclear force between quarks is brought about by interaction of virtual gluons. It describes their nature from the previous discussions – they are virtual waves without a core and they are obviously formed by background avyakta. ***Individual quarks contribute only about 1% of the total mass of the nucleus***, the rest is contributed by the 'mass-less' gluons when they are manifest, i.e., they have combined with the quarks. It is a proof of how much the background avyakta is directly involved in the formation and working of the world.

Experimentally, if the gluon could be stripped off the quark, it would show the corresponding antisymmetry of the quark that it was tied on to. Bombard the nucleus with extremely high energy particles; but to break two quarks held by a gluon would require so much force that it would amount to 'pulling out' a fresh antiquark through the gluon from the background avyakta. The pulling out of one quark leaves a corresponding volume of humble S plus or minus background toughness, and the defect is filled up by surrounding S in the reverse spiral and chirality as the quark is taken away, producing the anti-quark.

Level two of the strong force acts indirectly by transmitting the binding force of gluons through more 'virtual' particles such as pi and rho mesons.

Higgs field and Higgs boson: From the discussions about quantum vacuum, zero-point energy and virtual particles (the W boson), a rational idea about the Higgs field is also obtained. Avyakta provides a credible background for the Higgs field and also provides insight to how it works. The Higgs field is everywhere, and under the circumstances in which there is conductive background toughness, oscillations of avyakta impart toughness or humbleness which help manifest independent waves from the background or contribute to stabilizing their waveforms. The Higgs field represents one of the properties of the fabric of space, with the Higgs boson its scalar representation. But of late all the 'power' of the field has been attributed to the boson. – (*Strassler, M. (12 October 2012). "The Higgs FAQ 2.0". ProfMattStrassler.com. [Q] Why do particle physicists care so much about the Higgs particle? [A] Well, actually, they don't. What they really care about is the Higgs field, because it is so important.*)

1.8 The quantized wave

A unique feature of the universe is that everything is exchanged only in discrete packages. No loose sample is allowed. Also, all transactions are carried out in a move-stop-and-move-again style of action. 'Givers' move in jerks, 'takers' follow suit. It may seem rather odd but there is no exception.

If space were empty, energy could have been transferred without packaging. But here the entire space is taken up by the avyakta, therefore energy has to be conducted about through it, under its laws and using its units. Nothing is going anywhere on its own, because there is no 'space' for such activity. Even energy cannot enjoy 'freedom'.

What exactly would be the nature of transfer or conductance of energy between particles, through avyakta? It can only be in the shape of a wave. But background waves do not interact directly with particles. Only a wave which has 'particle' characteristics would be able to interact with the particle world as the situation would warrant.

The electromagnetic wave carries an oscillating unit of electric and magnetic fields in a single direction. Its job is transfer of that unit of energy between particles by *oscillating the fabric* in one direction at a corresponding frequency. In avyakta, the photon can be considered as a 'mini-particle', because it is formed as a result of an injection of a small quantum of oscillation from a particle into the fabric of space. Therefore it has nodal points for oscillation and is conducted linearly.

It is inferred that even the photon's true nature is complex, and it cannot be described in terms of a simple mechanical model (*Joos, George (1951). Theoretical Physics. London and Glasgow: Blackie and Son Limited. P. 679)*. But the known values provide indications to the components of its internal structure. The photon is a vector boson and has a spin equal to 1; apart from its oscillations, it has a spiral component, which gives a helical twist to the wave. In contrast to the wave form within the quark, the spiral here is not looped to itself to form a standing wave, but only imparts a helicity which allows for two polarization states.

The difference with the neutrino is that the photon does not carry any additional quantum of tough or humble S. Thus the photon has no antiparticle; the same is the force carrier in humble S.

To complete the picture of this kind of a wave, we may look at it also from the point of view of avyakta. S is made to suffer a disturbance to its composure when a small oscillation is injected into it. S The vigour of the effort by S to catch it is the same as that with which it 'flaps its wings' to speed away. Any frequency is possible depending on the energy conducted. Its speed in S therefore is the ultimate speed with which anything can ever move in that S. In other words, this speed is the index of the readiness with which the avyakta at that place can deal with a disturbance to its toughness – the vigour at that place.

To carry varying amounts of energy, it has a simple solution – the more energy it carries, the higher the frequency with which the oscillation is transmitted, i.e., the more number of oscillations per unit time. It makes sense; more energy will perturb or oscillate the fabric more. Only photons of a high enough frequency (above a certain *threshold* value) could knock an electron free. For example, photons of blue light had sufficient energy to free an electron from potassium, but even the brightest red light could not; as was found long ago in an experiment related to Einstein's solution of the photoelectric effect.

1.9 Antiparticle

This subsection follows logical inferences about antiparticles based on the discussions so far.

Assumption: In correlation with the symmetry as outlined in the Standard Model of particle physics, the fabric of space can host a set of both tough quantum and humble quantum, for a given range of background toughness.

Inferences: In a background of tough expanding fabric which is the phase of the present universe, a particle could be either a tough-in-tough one, a flat-in-tough one, or a humble-in-tough one. It can be inferred that the tough-in-tough particles are the neutron, proton and quarks, and the humble-in-tough particles are the antiproton, antineutron and antiquarks.

The tough-in-tough particle is formed when the fabric with extreme toughness spreads explosively as during that of the big bang. In the same way, it is assumed that the humble-in-humble particle forms when the fabric contracts after extreme humbleness (during the reverse phase of the universe). Thus it is a mirror image of the particle.

In the sun, the p-p chain involves formation of a diproton and its conversion to deuterium in the first two steps. Two protons combine, and one of them changes to a neutron. This would imply electron capture by the proton to convert itself to a neutron, which is the simultaneous process of all nuclei susceptible to β+ decay, especially proton rich nuclides. However, when there is more energy, like in the sun, the electron quantum can be assumed to be 'pulled in' from the surrounding avyakta to form the neutron, with reverse filling and the production of a positron. Such positron emission occurs in various β+ decays, but can take place only if the newly formed nucleus has sufficiently more energy that the original one. The neutrino emitted in either path is the same, because essentially it is the same process in both – electron capture and conversion of one up quark to a down quark.

Positron production occurs due to reverse filling as the electron is 'pulled in' from the background avyakta. So the chirality of the positron is opposite to that of the electron. Therefore the weak force is the only force directly involved in the chirality of the particle. The strong force is not involved in altering the internal spiral component.

The weak force interactions also present another very interesting phenomenon related to the chirality of the particle – only left handed fermions take part in weak interactions (and also the converse – only right handed antiparticle fermions take part in weak interaction). Our universe has a direct preference for left-handed chirality. Why? There is an answer – and it will be discussed in the section pertaining to the universe.

A proton and an electron are produced when the neutron relinquishes part of its toughness in beta decay. The proton is stable, and it can also make an arrangement with its erstwhile wave component for better resonance because the electron is flexible and can be adjusted to the surrounding toughness. The lack of a core in the electron is also ameliorated by adjustment with its corresponding tough original counterpart.

Since negative charge correlates with attraction for a lacking tough oscillation component, the humble antiproton has a negative charge; and since positive charge correlates with attraction for a lacking humble oscillation component, the flat-in-tough positron has a positive charge. Therefore the charges are reverse in the anti-particle.

1.10 Phonon, Particle, Photon and Fabric

What happens when any material is heated? The individual atoms are all wave forms; therefore they imbibe additional energy only with additional modes of wave function corresponding to the extra input. In a lattice this amounts to an additional oscillation to individual wave forms in a single frequency, or an extra mode of vibration of the particle. The additional oscillation is called a phonon. Within the lattice, the displacement of one or more atoms gives rise to a set of such oscillation waves through the lattice, with its amplitude determined by the displacement of individual atoms in the lattice. Acoustic phonons are produced by coherent movement of atoms in the lattice and are propagated in the form of sound. Sound waves in solids have transverse and longitudinal components, and phonons are regarded as bosons. The higher frequency phonons are felt as heat. In lattices in which atoms move out of phase with each other, eg, positive and negative ions, one moving to the right and the other moving to the left; optical phonons are produced. The frequency of the optical phonon corresponds to the inverse of the wavelength of the photon in vacuum.

An excited atom or excited nucleus returns to its ground state by releasing the difference in energy between the two states as a photon. The extra oscillation at the corresponding frequency is injected to the background and propagated as a point wave in avyakta. This spontaneous emission of photon also indirectly points out that all three - the mother particle, the photon, and the surrounding S – are one and the same. The ejected quantum is an extra oscillation of a standing wave, and they are both oscillations of the same fabric; they all appear different from each other by a simple yet definite set of rules that relate to the fabric of space and define our world.

SECTION 2

The Universe

2.1 Big bang and creation

From the four assumptions at the beginning of this presentation, particle creation can be looked afresh. Just before the big bang, avyakta is too tough to harbour any particle within it. Matter in any known form is non-existent. This means no gravitation, nuclear interactions or electromagnetic phenomenon. It is avyakta of near-infinite vigour.

When the expansion of avyakta sets in, the toughness begins to spread; and the natural outcome of the explosive expansion will be the creation of numerous waves in the fabric from the very small to the very big in the wake of the spiral unwinding. From this numerous units of spiral oscillations develop, and it is possible for some of them to become independent with stable resonance specifics - the earliest bosons, and later the pre-particles. This natural mode of creation is also evidenced by the fact that the same background avyakta can and does provide numerous virtual particle characteristics to support manifested matter waves in the present phase.

It is the unique behaviour of the core of the super spiral during its expansion that gives the impression that there is no 'explosion' taking place. The expanding spiral emanating from the super-tough core is almost explosive but it is 'invisible' till particles and radiation, begin to issue forth in its wake — almost like a jet plane flying too high to be seen or heard but announcing its course by a fabulous trail.

When a particle with an oscillating toughness comes into existence, gravitation would also manifest from the tough background. Thus the concept has easy and natural explanations for creation of matter and also the appearance of the gravitational force.

Corresponding anti-particles are not naturally created in this phase of the universe, except as break-down products. This is because the big bang is taking place in tough S, and it is far from humble to create vacuolization. Thus matter-antimatter asymmetry also has an easy explanation.

2.2 Generations of matter

The earliest subatomic particles would of course be quarks. But our quarks would not likely exist then, because the toughness of background S would not be conductive for them. It is reasonable to suppose that heavyweight quarks were the first. The symmetry of the particles and the corresponding slots as derived by the Standard Model depends on background toughness; therefore it can be assumed that in certain background toughness, the quarks with corresponding toughness (mass) are supported. As and when the background toughness gets reduced, quarks with lesser mass are favoured.

Are there corresponding heavy ones in the top rungs of the resonance ladder?

There certainly are. The heaviest quarks we can experimentally produce are the bottom and top quarks. They are unstable, produced in S of near infinite toughness that is impossible to duplicate in the present time. That background field would be incomparable to the present situation. But that part is obscure, and may forever remain so, because no force at our disposal would produce a semblance of that world again. High background toughness is not the same as high pressure and temperature used for collision in our laboratories. Anyway let us start with the quarks we know.

In fact, three rungs of the ladder are known to us. In our S, we have the up and down quarks which are constituents of our nucleus, and the electron. In the next heavy level we have two quarks, the charm and strange quarks, and a heavier electron, the muon. In the top most level we have the top and bottom quarks, and a very heavyweight counterpart of the electron, the tau (tauon). Coincidence?

The products of decay of the heaviest quarks are the intermediate ones; and the products of decay of the intermediate ones are ours. Again, coincidence? Possibly. But our simple neutrino has three faces – corresponding to the electron, muon and tau. Coincidence again and again? Very unlikely.

Particles in the other two heavier levels are not stable in our S. They can be produced experimentally in high energy collisions such as in particle accelerators which can provide some semblance to the high vigour and toughness of the universe near the time of the big bang. But in our vigour of S, they decay as soon as they are produced.

It is possible that the heaviest quarks decayed to the second generation ones without even forming a neutron; there must have been 'cascade particles' (that term can be understood in a broader sense now) too. But there is evidence that at least one heavyweight combination formed and ruled the universe in its earliest stage; otherwise the neutrino would not have a 'tau' face on it. It is possible that the second generation 'heavyweight' neutrons and protons would have lasted even longer (and even now as dark matter candidates in a background space of higher toughness). It is not possible to specify which of the quarks held hands to form the first super heavy proton. However in the next step (the second generation), a reasonable guess would be one strange, one down and one charm quark forming a heavy neutron, which, by one strange quark decay, would produce a 'proton' made of one up, one down and one charm quark (the lambda baryons).

The corresponding three quark and proton-neutron combination ratios (with the corresponding lepton) could have created a parallel world too. The second generation quarks have actually been observed in *hypernuclei*. (They are candidates for dark matter discussed later.).

The decay of proton is as such unknown now, because the proton is so stable now that such a question does not arise for a lot more billion years, till the toughness of S falls to that level. Proton decay can never be experimentally carried out in our part of the universe because, even if it is possible to induce particles similar to that in a tougher fabric by high energy collisions, there is no way to induce a situation akin to lower toughness of avyakta (humbler S).

It is a possible that all the quarks in the universe developed from a single type of great-grandfather quark, considering the isospin and symmetry. As the expanding avyakta spreads and reduces in toughness gradually the first heavyweight quarks would go out of resonance and develop a half-life. For the next range of toughness the next set of quarks are then favoured.

We call the process 'decay' but this term is a misnomer. It is through this process that the world was formed. Better terms to describe the process are splitting and differentiation. From the biggest base particles, all the corresponding slots would be filled downwards, indicating a simple master-plan; but a plan with a lot of ingenuity. Provide a background which behaves in a particular way, introduce the Big Spiral oscillation which would create the first particles; and everything is set. The playground is ready, the players are on their way, and the games would begin in their own sweet time.

Decay with resultant spectrum of quarks would bring up the ripe circumstances for the first nucleus which can be stable in that toughness. The virtual gluons manifest from the background field to combine with the quarks, and with the formation of electrons and protons, the charges become well established.

The prerequisite background condensations (bosons) are necessary for these changes to take place in the particle; therefore this precedes actual manifestation of the forces.

Thus a tough fabric of space in gradual expansion phase provides rational explanation for the generations of matter.

2.3 Formation of galaxy groups

As particles aggregate by gravitation to form larger bodies, those remaining for a time in the areas of high toughness at the arms of expanding wave due to the high gravitation and vigour may undergo gravitational collapse to form large black holes and structures like the quasars. The structures that form, with whichever association it has meanwhile developed, drifts with whatever velocity it has imbibed at birth or by association later, and they are either gradually brought together by gravitation and background oscillations of the fabric, or separated by the expanding fabric as its toughness falls. Stars form thermonuclear furnaces which fuse hydrogen and helium and form elements from carbon to iron, and then to still higher ones through various processes; while cosmic ray spallation produce lithium, berellium and boron. Later, explosive nucleosynthesis in supernovae produces very high elements. Very heavy weight elements could have also been formed in earlier supernovae when the S was tougher than the present.

Along with particles, the expanding wave also creates large scale oscillations. These perturbations are responsible for the subtle physics that result in the cosmic microwave background anisotropy. The reduction in toughness is also not perfectly identical owing to the nature of the expansion and the inherent oscillatory nature of avyakta, and some regions may remain tougher and others may become comparatively humbler, as is observed in the distribution of later galaxies. Oscillations and gravitation bring matter together, and it condenses in intricate web-like structures based on the background fabric, forming galaxy groups and clusters.

The present concept in Physics lacks the description of the fabric of space; therefore the reason for these oscillations and structural condensations are not explained at present.

2.4 The explanation of dark matter

The high vigour and toughness of avyakta explains faster accretion and formation of structures in observed patterns; however this cannot be explained if space is assumed 'empty' and only known particles (present stage of quarks) are considered in the universe. Therefore dark matter of unknown nature was postulated to add more mass, to account for higher gravitation and observed phenomena. It is now believed that dark energy plus dark matter constitute 95.1% of total mass-energy content in the universe. The proposed dark matter has similar characteristics of the tougher background avyakta such as – it does not react with light or any other particle directly, it is hidden and only interacts through gravitation.

Puzzle 1: Mass of galaxies were calculated from the distribution of stars in spirals and mass-to-light ratios, and it was found that the stars, and indeed whole galaxies rotate with more speed than warranted by their total mass. They should go a lot slower if they adhere to the law of gravity - otherwise they should have more mass due to dark matter presence.

Explanation: Earliest galaxies were mostly composed of gas and had only a few stars. They gain mass when small galaxies join together, but when the total mass becomes high enough and covers a large area, gravitational accretion begins to curve the surrounding fabric. The basic mode of oscillation of avyakta is spiral; and the fabric owing to high toughness spirals inside. The high gravitation at the centre along with the incoming spiral of toughness leads to gravitational collapse and a black hole at the centre, around which a spiral precession is formed. The spiral of a stable galaxy has many similarities to an oscillation within a particle, though here the spiral is not looped. The precession of the incoming spiral and thus the major part of its energy is due to the high toughness of the surrounding avyakta. But this background vigour of the fabric of space is not considered at present; only the mass of the particles within the galaxy is taken into account. Also not considered is the presence of other dark matter candidates in areas of higher toughness. (Since space is considered uniformly empty, it is assumed now that heavy matter such as quarks of the previous generation will be as unstable everywhere in the universe as they are here.). Hence the extra mass of unknown dark matter is needed to explain the phenomenon.

Puzzle 2: The orbital velocity of planets in solar systems decline with distance. But the orbital speed of stars within the galaxies does not follow this pattern; stars revolve around their galaxy's centre at equal or increasing speed over a large range of distances, i.e., they more or less maintain the spiral pattern.

Explanation: Present solution to this puzzle is to assume the existence of dark matter and that too distributed in an exact way, extending from the galaxy's centre to its halo, in a roughly spherically symmetric pattern. The 'density wave theory' proposed by C.C. Lin and Frank Shu introduces the idea of long-lived quasistatic density waves– that the arms of the spiral are made of greater 'density' in the form of 'waves' which actually maintain the spiral pattern. It confirms to the spiral oscillation of avyakta, where the spiral wave itself cannot be observed but is followed by the stars in that background. The pattern of the spiral is maintained by avyakta but in present physics only the particle world is taken into consideration, therefore dark matter in the specific pattern was required for explanation.

Puzzle 3: In galaxy clusters, gravitational lensing observations confirm the presence of considerably more mass than is indicated by the clusters' light, so dark matter is considered to account for it.

Explanation: There is considerably more toughness involved in the galaxy than its surroundings, due to gravitational collapse at the centre plus the incoming spiral, but only mass from the particle world is taken into account at present, and that too corresponding to particles in background toughness in our part of the universe. Lensing depends on incoming gradient of toughness and this corresponds to curvature of the background fabric due to higher toughness in the galaxy.

All the reasons for dark matter and even the presence of dark matter candidates are easily explained by considering the changes in consistency of the background avyakta which interacts dynamically with the particle world. For example, the high toughness within the neutron stars would 'reverse-produce' the strange and charm quarks from the present neutrons, which leads to magnetic field breakdown and production of ultra-high energy cosmic rays. These are named 'strangelets', and they are also suspected dark matter candidates, and they confirm to the model of avyakta. Present physics also does not know how these ultra-high energy cosmic rays are produced.

A similar picture can be seen when galaxies grow old, their stars becoming few and redder with all the gas being used up. There is considerable amount of dust and more black holes. The spiral implodes more and goes out of pattern resulting in more toughness concentrated within it, forming 'darker' galaxies. There are galaxies in space having much toughness contained within it but too few stars to account for the mass by themselves, for example the dwarf spheroidal galaxies. Their dark matter is made of high toughness of avyakta plus heavier matter in it; matter which cannot exist in our part of space (and which cannot be explained by present physics owing to lack of the concept of a background fabric which can undergo change in its consistency.)

The large scale structure of the universe has many 'void' spaces, 'nodes', 'filaments' and even 'walls'. Immense voids create spaces, sometimes called 'cosmic web'. Structure that cannot 'normally' form and 'cold space' have no explanation under the present cosmological model. All these can be explained only by avyakta and its pliability which provides variable tough and humble features thereby making different laws possible in different parts of the universe.

There is feedback influence from the particle world to the surrounding avyakta. Increase in density of particulate matter by gravitational accretion in heavenly bodies in turn increases toughness of avyakta involved in that body. The effect is pronounced in large stars, galaxies and in the formation of black holes.

Gravitational collapse means formation of a super tough node of the fabric of space going to contraction phase. In the black hole, waves of contraction of avyakta pull everything in its vicinity to inside, dragging them around the node of ongoing contraction. The effect of the black hole becomes pronounced as a photon gets closer to it. If the path of the 'ray' is close enough and/or the conglomeration it passes adequately large, the ray takes a smooth and inward bend, the reason for gravitational 'lensing'.

In future most of the matter which provides the toughness inside the black hole is converted to Hawking radiation. Infinite toughness or infinite humbleness are the phases in which matter is dissolved back to the avyakta. This indicates that there is loss of information, but there is also every possibility that all information will be recreated, maybe in a different pattern, as universe renews with each expansion phase (big bang), assuming it to be cyclical.

Without matter in high gravitation which was the cause of the black hole, the contraction point in the background avyakta is no longer sustainable and this phase ends. The point of super tough fabric slowly dissolves as a background wave which cannot be observed by any instrument, gradually evaporating the node. Black holes and galaxies are random events; not a stabilized oscillation like the pulse of the universe.

2.5 Red-shift

So, man and his abode are HERE, somewhere on the way behind the onslaught of the explosive expansion. The vigour of S hereabout is on the decline even ‘now’. We, from a planet in space, look out to see a lot of structures in space. The further we look, the further we see the past, as light gets so many years to get here. We see galaxies that have long ago crumpled to dust. But we also see that they are all running away from us. The universe is expanding. The speed of recession is calculated by red-shift, and the further they are, the greater the shift.

But avyakta was not considered before in explaining the red-shift. It has undergone expansion from the big bang onwards. Therefore the galaxies are not running away; as the background expands the objects in it are getting gradually separated from each other. This is the reason for the red-shift.

Our electron 'sees' the same wavelength for a wide range of toughness of the background in the same fashion as it also spreads along with the surrounding toughness (Similar to the photon). But in what fashion does it 'see' visible light that originated even before it was formed, for example near the beginning of the universe? That photon came from a toughness which was many times the background toughness conductive to our electron, and it has spread out that much. It is possible that the cosmic microwave radiation could have a reason other than heating of the universe. In the phase of creation in which all pre-particles were manifested, a tremendous amount of radiation had also been created. Our electron would not 'perceive' that photon in the original region of the spectrum.

2.6 Possibilities in the far future

As the wave of expansion of avyakta crosses the expansion phase, universe would become ‘humble’, i.e., its toughness reaches unity and then goes below it. As it reaches the flat stage, gravitation ceases because the tendency of avyakta to expand would be present no more. As it goes humble, S would develop the tendency to shrink. The gravitational wave can start to repel – formation of the anti-gravity wave.

At the end of expansion phase, avyakta would have become infinitely humble; the particles would continue its stepwise decay (including proton decay) until it is dissolved in humbleness.

Once the contraction starts, the resultant turbulence would

bring up the humble-in-humble particles from the infinitely humble core; the true antiparticles, instead of the tough-in-tough particles created during the Big Spiral Bang. There are no stars, no planets in this reverse phase, because gravitation repels; it is eternal night which ends as the contraction phase reaches culmination. The anti-particles decay and fade away as avyakta regains toughness, and the universe is cleaned-up for the next Spiral Bang, when everything is freshly created. On its own the universe renews its interior décor at every cycle of oscillation.

In a background of humble S, mirror images of both our particles and antiparticles are formed. The humble-in-humble antiparticle is different from the humble-in-tough antiparticle in that its content is even 'less' than our antiparticle. That means that the 'antiparticleness' here is applicable only to this phase of tough S; the real mirrors are even more radical in that the whole picture is reflected and not just the particle. It time can be perceived as running back, we can safely conclude that CPT symmetry is truly preserved.

All the particles formed in tough S (our phase) have one thing in common: The direction of the spiral of toughness in them (chirality). It is obtained from the super tough 'super-spiral' of the big bang. In regions of tough S the spread of toughness within them and emanating from them always spirals in this direction. In the reverse phase of humble S these directions are of course mirror images. So is our universe left handed or right handed? By now those who are familiar with decay modes would have already guessed the answer – our phase is left-handed. And the mirror universe in the reverse phase is of opposite chirality to this phase.

2.7 History and philosophical implications

In earlier days when it was understood that electromagnetic waves needed a medium for propagation, a medium called 'aether' was the accepted solution. Fields were visualized to be based on this medium. When it was proved that every particle of matter had wave characteristics, it indicated that the particle and medium were not two but one, and the particle moved through the medium by transferring (waving) its characteristics through the fabric of the medium. But instead of reaching that obvious assumption, it was maintained that the medium was different from the particle. It was even believed that this medium, the aether, moved like a 'wind' around the heavenly bodies, separate from them. The famous Michelson-Morley experiment gave conclusive evidence that such aether could not exist. (Bodies moving in it would experience a wind of it. The velocity of electromagnetic waves measured in this direction would therefore be more than their velocity in a direction perpendicular to that. In 1887 a very careful experiment to test this was carried out at the Case School of Applied Sciences in Cleveland by Albert Michelson and Edward Morley. They found that the velocities of light in all directions remained the same.) After some desperate measures to salvage aether, and again not seeing the obvious, the world of physics took the opposite view that there was nothing there whatsoever; space was 'empty' to the core, a 'vacuum'.

Under that prevalent idea, Albert Einstein's solution was to discard the notion of a medium at absolute state of rest (aether). In relativity, any reference frame moving with uniform motion will observe the same laws of physics. In every reference frame, Einstein took the speed of light as a constant (without the background, some base has to be there for all the other values to relate on), and brought about the concept of 'space-time', fixing any event to be happening at a time and a point of space. Two objects will have two different sets of reference frames and they can be described, by (t,x,y and z); and (t1,x1, y1 and z1) where t is time and the other three are the co-ordinates in space. Then the Lorentz transformation is applied to derive the relationship between time, length, and mass change for an object while that object is moving relative to the other, keeping speed of light a constant.

Since the speed of light is the speed of any wave in the model of avyakta, and other values are all 'observed' and relative values, from the equation of the Lorentz factor we can infer that such a comparison would explain the characteristics of the background fabric itself.

Ironically, Lorentz transformation was not fully Einstein's; it was derived from the Lorentz-Fitzgerald contraction hypothesis, which was put forward to help explain one of the properties of aether! Length contraction was postulated by George FitzGerald (1889) and Hendrix Anton Lorentz (1892) to explain away the negative outcome of the Michelson-Morley experiment and rescue the hypothesis of the stationary aether. Hendrix Lorentz was one of the staunchest supporters of the aether theory.

Relativity is correlated with the concept of avyakta easily. For example, what happens to the particle when it is accelerated to speeds very close to that of light? 1). The particle cannot move above the speed of light because that is the maximum speed at which any wave can go in that particular background consistency of the fabric. Therefore as its speed approaches maximum limit, the energy required for accelerating it, or in other words its mass tends to become infinite. 2). When the velocity reaches near to that of light, internal oscillation in the direction of motion will get shortened because it cannot move above that speed; in other words it will lose its third dimension - the phenomenon of length-contraction. 3). The total speed of a wave cannot exceed the speed of light. A particle can move near that velocity only by its wave components compensating the speed of its other oscillations. In other words, the speed of its internal beat will reduce proportional to the speed of its extra vector component. Therefore its internal clock slows down – the phenomenon of time dilation.

What Einstein did was to brilliantly work out a way to advance physics ignoring the concept of the 'background', after the failure of the 'aether theory'. Whether the background was empty or not did not matter anymore at that time. Therefore the aether theory was not needed and soon fell out of favour.

Plotting events relative to each other in three co-ordinates would enable the geometry to be studied without needing the background. Making time the fourth co-ordinate only means that an event would occur at a point of time, it does not explain why or how. But adding the time co-ordinate would bring a pattern to the geometry as the Minkowski space-time, based on which models could be explained, mathematically advanced and even predicted. This model is different from the actual four-dimensional space.

The Standard Model of particle physics was also built up based on the same concept and principles, in which photons and other elementary particles are described as a 'necessary consequence' of physical laws having certain symmetry at every point in space-time. It ignores the background but explains the symmetry in it. Three of the quarks combine in 1:2 ratios to produce either a neutron or a proton; both the products have similar masses. The free proton is stable, the free neutron is not; they combine in certain known ratio to produce elements. From observations like these, the conclusion was arrived that there is a pattern and a symmetry regarding the structure of these particles. Mathematical formulations of the symmetry of subatomic particles was similar to mathematical formulations given to spin, therefore it was named isospin. Later this was broadened to a large group, the set of 'flavor symmetry', and there was found to be a similarity about how these particles related to each other with the representation theory of SU(3) symmetry. The properties of the various particles can be related to representations of Lie algebras, corresponding to "approximate symmetries" in the universe, and the different quantum states of an elementary particle give rise to an irreducible representation of the Poincaré group. Thus even with the double-blindfold of obscured physical nature of the particles and the ignored background, the Standard Model of Particle Physics was successfully developed. Physical measurements, equations and predictions from them are all consistent and hold a very high level of confirmation.

Even after it was confirmed that gravitation 'travelled' with the speed of light (definitely indicating a medium through which it propagated), the notion of 'background emptiness' prevailed. 'Warping', folding or curving of space-time also did not fully erase reservations about 'empty' space transmitting wave packets. Moreover, how does 'nothing' warp, curve and fold on itself?

After undergoing a laborious 'renormalization', quantum electrodynamics became the most successful theory with experimental data confirming its predictions to such a high precision; so that the simplest and most basic questions were forgotten and they remained unanswered. As Feynman remarked about renormalization, "there is no theory that adequately explains these numbers. We use the numbers in all our theories, but we don't understand them – what they are, or where they come from. I believe that from a fundamental point of view, this is a very interesting and serious problem."

The phenomenon of neutrino oscillation and virtual particles give further evidence that matter is not separate from the background. Wave-matter duality and the fact that the all matter is represented as waves at the quantum level also points to direct evidence of a background avyakta which is not separate from matter (A separate medium like aether cannot explain this fundamental property of matter.). But fields were perceived out of nothing and energy as stored in empty space. Proposed were the ideas of inflaton and Higgs fields and numerous virtual particles in empty space, and it was argued that mass-less particles would suddenly develop mass. While virtual particles were used to explain everything pertaining to the three fundamental forces and the interactions at the subatomic level, hypothetical dark matter was depended on to explain the large-scale structure of the universe.

Many physicists have become resigned to the foregone conclusion very early in their career that it didn't matter if you don't understand it; you just have to prove it mathematically. Yes, mathematics is the language of the universe. ("Mathematics takes us into the region of absolute necessity, to which not only the actual word, but every possible word, must conform." - Bertrand Russell.) It is clear cut, irrefutable logic and absolute evidence; but of late it is forgotten that mathematics is the way and not the end. For example, are we doing the right thing by accurately calculating that there is a valley full of beautiful flowers exactly 658 meters ahead; arriving at the complete geometry of the flowers and comparing this valley to other valleys to see how they change with time; when we have blindfolded ourselves and won't see a thing? Why the foregone conclusion that the physical aspect of it is too complex to comprehend? Since when did physicists start giving up so easily?

If the size of individual atoms is compared to the size of a football field, the nucleus is only about the size of the football. Therefore most of matter is perceived as 'empty' in the present view. And it does not end there; the nucleus itself is also 99% virtual; only 1% of it is contributed by the quarks. If physics discovers some way in future which can provide a peek inside the quark, it would find 99% of the quark to be 'virtual' or 'empty' too because it is also nothing but a wave form. We can then even visualize ourselves as 'completely empty' men! As long as there is no concept of the background avyakta, the whole picture would continue to be interpreted inaccurately. Actually none of these spaces are empty, they are all occupied by avyakta, all matter manifesting as waveforms of it.

But to grasp the concept, physics has two practical problems to overcome: 1). Acceptance of the idea of the underlying fabric of space will involve re-structuring the edifices built so far. 2). It involves accommodating an entity - avyakta - that cannot be weighed, dissected or transformed with any present device, as any equipment is necessarily its product and changes itself as and when the background changes. This is also the reason why we are unable to provide direct experimental evidence for avyakta. Conclusions regarding physical phenomena are also derived from direct observations, but avyakta is beyond the observational capacity of the sense organs. There are only indirect evidences such as the gravitational wave, but the number of such pointers is increasing along with advancements in physics.

And that is precisely the basis of this endeavour – the attempt at a simple set of explanations from obvious evidences. All the entire enigmas in physics can be explained by the concept of the fabric of space instead of the round-about way resorted to till now. Every finding will hold good, all mathematical derivations and observations in relativity would remain the same; only the background concept needs to change.

PART 2

Philosophy

TOWARDS CONVERGENCE

Introduction to core Vedantic concepts

Cosmology in particular and physics in general can overcome many an obstacle these disciplines confront now if the concept of the three-tier universe envisaged by ancient Indian philosophy is given a try. This philosophy is loosely termed 'Vedanta' and does not represent any religious outlook but rather a well thought out rational insight about the secrets behind the working of the universe. The word 'Vedanta' means 'at the end off the Vedas' referring to ideas generally mentioned in the Upanishads.

At present there are no standard texts or works that deals exclusively with Vedantic thought. There would have been once. However ideas from this earliest knowledge had been prevalent in ancient India during the time of the Upanishads and Bhagavad Gita, and therefore the core concepts of this almost-lost wisdom has found its way into these verses. It is to be noted here that in describing these, a religious study is not being attempted.

Physics cannot provide idea about the force behind *avyakta* but Vedanta can. This is because the methodology of the former is different from the latter – ancient Indian philosophy makes logical assumptions and then relies on rational thought processes for corroboration. Lack of mathematical or experimental evidence doesn't matter, only that all physical phenomena must confirm and should not contradict it's assumptions in any manner, and its predictions must come true. Therefore this style of reasoning is adopted when the going becomes non-computational or un-quantifiable. For example, 'love'. This phenomenon is highly subjective and varying, therefore experimental evidence is bound to be sketchy and inconsistent. It is non-quantifiable and so it cannot be derived mathematically. It is also incomparable among specimens. Same is the case with 'goodness', 'truth', 'happiness' etc.

Though Vedanta is generally considered very tough to grasp, it is in fact a simple concept. Its central idea can be accomplished through logical answers to four easy questions. Question one: Isn't it logical to assume that there is a unified force forever basic behind our vast, complex and ever changing universe? The obvious answer is yes. Question two: Where does that force reside – in some nook or corner of the universe or everywhere at the same time? The latter would be the logical answer. Question three: As we are in the universe, aren't we too part of that force? Again, the answer is yes. And the last question: Whatever else we are – body, mind and intellect – being perishable, isn't this factor the only permanent thing in us? Again, simple logic demands a positive response. The moment this question is answered in the affirmative we arrive at the first of the four 'great statements' (*mahavakyas*) of Vedanta, namely 'Tat twam asi' ('That is you').

The next step in the study of Vedanta is to learn how this fundamental force functions. It is invisible and un-manifest, therefore there must be a pattern of working behind the universe which is hidden. So although the entire universe is this force's manifestation, Vedanta considers it to have three facets. One is the observable universe, ever changing itself. This is called *Kshara* (perishable). The other is the unseen template on which the *Kshara* is based, and also comes from and goes back to. It is named *Akshara* (imperishable). The third is the factor called '*Isvara*' (The force behind all this phenomena, commonly called God, for whom many religions in the past have attempted to provide various attributes in the past.). Everything in the universe including man exists in all the three at the same time – the perishable, the in-between template and the third entity.

Discussions

Even though the basic idea of the fundamental force is simple, the three levels need adequate discussion for lucidity. It is therefore prudent to research a few verses that represent core Vedantic ideas from the Gita and the Upanishads:-

It is said that there are two purusha dwelling in all bodies in the universe. 1.- Kshara - that which is perishable, which is the form or structure or manifestation. 2.- Akshara - that which is imperishable; that which gives rise to all structure or manifestation, the kutastha.

Bhagavad Gita 15:16

(*Purusha* = entity, Kshara = that which is perishable, *akshara* = that which is forever, *kutastha* = like an anvil of the black-smith, which allows every body to change its shape but itself remains changeless.)

How can empty space be the creator of all that occupies it? Therefore there must be a greater reality even beyond space and it must be ever-present in equal measure everywhere. That is *akshara*, the substrate of dialectical nature.

Every structure and manifestation in this universe exists in all the three levels at the same time. Humans experience one – the *kshara* world in which all these creations are born, exist and die, our own bodies included. The other level manifests this world but remains forever unaltered and unobservable by any physical means. The former is known as *Kshara* meaning that which perishes and the latter as *Akshara* meaning imperishable. *Akshara* is like the anvil on which many a metal object is shaped but remains unaffected by all the forces it is subjected to. (*Purusha* generally means an 'entity' and has nothing to do with gender.)

This unobservable fabric of space – the *Akshara* or *avyakta* – pervades the entire universe. Matter is just a wave oscillation in the fabric of space or *avyakta*, harmoniously syncing its frequency with the *avyakta* around it, moving in *avyakta* by transferring its content (a quantum of energy) from point to point.

Physics needs constants to settle equations; it depends on the invariance of units. But this background template cannot be measured, weighed, dissected or handled with any device, as any equipment is necessarily its product. There is no direct scope for experimental evidence. And this ancient philosophy makes it all the tougher to physics as it completes the picture by adding yet another level of reality to the two already pointed out:-

The third or Supreme purusha, distinct from the other two (the Akshara and Kshara), is called the Paramatma (the Highest Self), the One pervading all the worlds, the Ruler, the One who does not change or decrease, Isvara (God).

Bhagavad Gita 15:17

A force induces *akshara* to pulsate; that, therefore, must be the most fundamental and omnipresent. *Paramatma* or *Isvara* or God or whatever one may call It is the basis for the universe. It doesn't assume a name because there is no other like It and also nothing separate from It; though men have named in it in many ways. *Akshara* is Its dialectical entity in which It continues to be imperturbably present. It gives *Akshara* the impulse to pulsate, making it the fulcrum balancing upon which myriads of small and big pulsations come into being, exist, interact and get dissolved.

That which is not eternal has no true nature of existence. That which is eternal never ceases to be. These two states have been well

discerned by the wise who could envision fundamental truth – the tatwadarsis.

Bhagavad Gita 2:16

(Tatwadarsi is one who has fathomed the secret the state of *brahma*; *tat* means *brahma*.)

This verse represents the major difference in present scientific concepts from Vedantic style of thought. Present scientific view fights shy of considering an Ultimate Cause because then it should have another cause behind it! It refuses to see the Ultimate Reason behind the apparently endless chain of cause effect relationships.

But Vedanta had long ago resolved this issue. The distinction was easy. Anything that forms and changes with time cannot be eternal. That which is eternal is not born or formed, nor does it undergo change or death. This verse indicates the beginning of a thought process which led to the realization of the stable and unchanging basic framework for the material world and outlining of the three levels of the universe.

What is to be reckoned as indestructible is that by which everything in the universe is pervaded. None ever can cause damage to that, the indestructible.

Bhagavad Gita 2:17

In continuation to the previous verse, two important characteristics of that Ultimate Start for all cause-effect relationships are understood. One: It should be imperishable. Two: It should pervade everything and every place in the universe. Therefore logically, one may comprehend it, but it cannot be observed because it pervades everything including any machinery of observation, measurement or destruction. This is named *Akshara* meaning imperishable, the dialectical substrate from which all cause-effect relationships and action chains start.

But there is a different presence even beyond akshara; it is the secret of secrets and eternal; it is not wiped out even when the entire set of creation is destroyed.

Bhagavad Gita 8:20

This verse exemplifies the happy progress of the one who strives to know the Supreme Entity through his intellect as these ancient researchers had done. First, the temporary and continuously changing character of the obvious world (the *kshara*) with its perishing and renewing nature gets well understood through constant observation. This leads one to the understanding of *akshara* the ever-present substrate from which the material world is born. Its discovery is based on the obvious conclusion that if something huge like the observable world could exist and change and renew like this, it must have a stable base to do so. This contemplation and meditation unravels *avyakta*, the basic fabric of space. Then, life in its plethora of beings with their patterns of birth and death are observed and contemplated on. It is then realized that there must be something greater than a mere mechanical substrate, which forms the basis of all life, and that which is beyond all changes in the universe making such manifest life steadily possible. It is then recognized as the 'secret of secrets'. Something that is even basic to the *akshara* would point to the most hidden but ultimate truth.

I am beyond the kshara, and also greater than the akshara. So in the world, I am denoted as Purushottama (Supreme purusha) in the Vedas.

Bhagavad Gita 15:18

The logical reason for absolute greatness of that Highest Entity is: it is the third level and the highest reality and permeates the entire universe. Those among men who really know 'me' therefore calls me *Purushottama* (the noblest *purusha* or entity); so also, all sciences that seek the basis of the universe too.

Modern science has been engaged in an all-out effort to unify the fundamental forces. Till a little while ago they were four in number: gravitation, electromagnetic, weak-nuclear and strong-nuclear. The second and the third have of late been combined as electro-weak. There are efforts to reduce the number further and finally pack everything into one but this cannot progress without recognizing the fundamental forces as different facets of the same force (*Purushottama*) expressed through the dialectical nature of *akshara*. But science has no direct material evidence to recognize any level of reality beyond that of the perishable world. The Ultimate substrate for all cause-effect relationships (*Akshara*) is taboo; and God is heresy.

From the insight in the previous verses, these are logical conclusions to such a substrate that hosts everything. Physics already has numerous indications and clues to the background fabric of space. It has also understood that there is a background energy in 'space' and has visualized numerous virtual particles arising out of it. The fact that all matter has wave component is itself definite proof of the unity of space-matter manifestations. But the concept of a mother substrate in which all matter and space is based on is yet to be fully explored. 'Space-time' and the 'curvature' of 'empty' space regarded as the reason for gravitational 'waves' do not give full credit to *avyakta*.

Why is this invisible background not evident to physics?

This is Avyakta (beyond the observational capacity of the sense organs), incomprehensible (beyond imagination) and cannot be

understood by changing (beyond transitions of form and content). Know it as such and give up all misery.

Bhagavad Gita 2:25

What prevents physics from directly visualizing *avyakta* is discussed here. In science, conclusions regarding physical phenomena are derived from observations made through sense organs. But the fabric of space (*avyakta*) is beyond the observational capacity of the sense organs. This is because all sense organs are its manifestations; so *avyakta* cannot be seen, heard or touched. The term *'avyakta'* in Sanskrit means not manifest; unknown quantity or number, primordial element or productive principle whence all the phenomena of the material world are developed. It describes beautifully physics' woes in grasping it. It cannot be quantified mathematically because it is continuous and therefore no instrument can be used to quantify it. Experimentation is another method science employs to know an object. Here, the object under study should interact with physical agents and/or transform itself with time. But *akshara* is not reversely modified by its products.

This 'all-pervading fabric' behind the apparent emptiness supports all physical phenomena that happen in the universe. *Avyakta* encompasses the universe as a continuous template. It is everywhere in the universe. It can in no way be manipulated or changed by any 'visible' or 'manifest' object or radiation in the universe because then it would disqualify for the universal template.

It is a master medium. It hosts solid, liquids and gases; therefore it cannot be classified into any of them. It can undergo infinite homogeneous contraction followed by expansion, as in the big bang, indicating gaseous character. Yet it has solid characteristics. For example, it was deduced long back that an elastic medium propagating electromagnetic radiation at the speed of light would have to be so tough that no planet or body could ever twist, turn or wiggle in it. It can also host the toughest of solids, and the particle that it hosts is in wave form but does not even out like water or gas. Yet gravitational waves are conducted through it very possibly just like waves in water, indicating liquid character.

Even the wave form of the particles that it hosts is so highly complex that physics is unable to deduce the true physical nature of the particle. *Avyakta* holds black holes and supernovae explosions within it without the slightest harm to itself. It is beyond the grasping power of the mind, because all its physical characteristics can never be deduced, and the full nature of it is bound to remain unknown, unknowable.

This vast Brahma is the birthplace of My creation. In That I seed. Thus everything here happens or manifests.

Bhagavad Gita 14:3

The fundamental force – the *Purushottama* – deposits the seed of pulsation in *brahma* (which is another name for *akshara*), resulting in the primordial pulsation from which the manifest universe is born. *Brahma* literally means 'that which has the tendency to expand'. The mode of expansion is spiral. The fabric of space within the expansion volume is *avyakta,* from which all mater and radiation are manifested. (The similarity of these insights with the Big-bang concept is obvious.) The 'tendency' to expand is maximum at the most condensed stage when *prakriti (brahma)* is most withdrawn into *purusha (Purushottama).* At the pinnacle of condensation the seed of the tendency to expand sprouts. *Purusha* and *prakriti* are One before the seeding; both are equally real too. The primordial pulsation is the cause for the manifest world.

For whatever manifestations or forms taking birth, in any place

of birth, the vast Brahma is the true birthplace, and I am the Seeding Father.

Bhagavad Gita 14:4

Just like the pulsation that is the universe, the purposeful energy seeded into every subatomic particle during its birth has three tendencies (*gunas*) in its oscillation that make it sustain itself and also enables it interact with its kin. They are the expansion (*rajas*) and contraction (*tamas*) phases in oscillation and the tendency to remain stable (*sattva*). Verses pertaining to these *gunas* are to be seen later. Variety among them arises on two grounds: one, the *content* of energy in the oscillation, and two, the state of *avyakta* around it. The consistency of *avyakta* at any point anywhere in the universe keeps on changing based on its expanding nature of this phase of the universe. Unifications, dissociations, leakages, grabs, fissions and fusions go on relentlessly as long as *avyakta* is neither too condensed nor too expanded to form or harbour any oscillation – at the pinnacles of the two phases during its own oscillation.

The day of brahma lasts thousands of yugas (eons), and then for thousands of eons it is night. People who know of them know true day and night.

Bhagavad Gita 8:17

'Brahma Deva' is the concept of a creator-god based on this cyclical nature of pulsation of *brahma*. As the material universe is created from *brahma*, the Brahma Deva is visualized as being born when the Supreme seeds *akshara* with the urge to pulsate. In a unique symbolization, Lord Vishnu representing the Supreme Entity lies in yogic sleep on Ananta, the snake. Ananta (literal meaning 'the endless') curls his body in long loops one upon another to provide a unique reclining sofa for lord Vishnu and spreads its hood with the thousand heads in an array forming an umbrella above. Lord Brahma sits on the lotus flower stemming from the midriff of Vishnu.

The visual symbolizes the beginning of the universe. Endless time lies coiled on itself. On it the Supreme lies in yogic sleep, sublimely quiet. The universe begins to bud and blossom from its midriff. On that blossom manifests Brahma Deva, the creator with four faces; representing *brahma (akshara)* from which the material universe (*kshara*) is created. (There is a possibility that the pictorial representation indicates that *avyakta* is four dimensional, or it may simply indicate four directions.)

The universe itself as well as every oscillatory wave in it has its own unique frequency. Time can be reckoned on the basis of any of these. But as every oscillation the universe contains severally dissolves back into *akshara*, such times too will cease to be. Ultimate time at the level of the third reality is timelessness. One cannot measure it with 'knowable' time, it is incomparable.

The ancient Vedantic thinkers also understood that any reckoning of time apparent is subject to changes in the constitution and character of the medium of *akshara* brought about by the act of pulsation of the entire universe, as the frequencies of oscillation of wave components of all matter depend on the consistency of *akshara*.

In the morning of that day, from that which is called by the name Avyakta, everything becomes manifest. When night comes, into that Avyakta itself, everything dissolves or fades totally.

Bhagavad Gita 8:18

The beginning of expansion phase of the universe leads to the formation of numerous oscillatory waves in *avyakta*. All matter is no more than condensed packages of energy oscillating the same way the universe does, the energy inside condensing and expanding spirally. The physical world – *kshara* – is the outcome of such waves moving, interacting with each other and coalescing in an expanding ocean of energy.

After the expansion cycle is done comes the reverse phase and all oscillations are by then dissolved back into the fabric. The contraction cycle is night of the *brahma*, which hosts the world of mirror particles, and no star or galaxy can exist in it because of antigravity – total darkness prevails. This phase too would end, and the universe is wiped clean for the next cycle.

As the spider spins out the web from its stomach, as the hairs grow from the human body, as the plants sprout from the earth; so does the universe manifest from the akshara.

Mundaka Upanishad 1.1.6

One of the specialities of Vedantic methods is avoidance of unscientific or miraculous representations for explanation. Expression of the third level, the *kshara* world, is natural, and here such accustomed examples are given to grasp the concept. The verse is exemplified in both poetic beauty and meaning.

This is the truth. In the same way as the numerous thousands of bright sparks fly from fire that burns well, numerous worlds originate from this one akshara, O respectful student. All those worlds that have manifested from this akshara dissolve back in itself with time.

Mundaka Upanishad 2.1.1

The entire set of creation (in this universe) is manifested again and again during the day, then are spent and dissolved back at the advent of the night.

Bhagavad Gita 8:19

The universe and its numerous creations are manifested from *avyakta* during each cycle. These particles and bodies become stable, and carrying out their numerous roles and parts they have to play, each becoming a component of the colossal epic of the universe. Then that phase is done, and they are dissolved. After an equal epoch of silence and darkness, again the next cycle comes. Exact replication of creations of the previous cycle is neither necessary nor warranted.

All creations have their beginning in Avyakta, become vyakta (manifest to the five senses) for a while in between, and again become one with avyakta when they cease to be. What is there in it to grieve?

Bhagavad Gita 2:28

Eternal reality is the fundamental force, and its own dialectic form, the non-manifest substrate, *akshara*. In the light of that, the ever changing apparent universe is like a dream. But the present world of science is stuck with just that. The belief that there is only this level for the universe, and for us too, makes us attached to the perishing nature of this world, because everything in this world has got an end and death is a certainty. But when one comprehends the immortal levels and his own true nature, there would remain no more reason to grieve about anything. Vedantic thought is clear that our life force stems from that One Entity, and knowledge about the three levels is the first step that this ancient philosophy considers for being one with that entity.

My prakriti (obvious nature) is split into eight: bhumi, jalam, vayu, agni, akasa, mind, intellect and ahamkara.

Know that this obvious nature is apara. The nature which is distinct from this, in which all life and entities in this universe is based on, is My para nature.

Bhagavad Gita 7:4, 7:5

Prakrti means nature. There are two types of *prakriti*. The *apara prakriti* is the nature of the *kshara* level – the ever changing world where there is both origin and destruction – the world that we perceive. This is the world dealt with in present physics.

Para prakriti is the nature of *akshara*, the substrate or medium from which *kshara* is born, supported on and gets dissolved back to, and it acts by its nature of resonance through *avyakta* (the fabric of space) reflecting its dialectical nature to the *kshara* world.

In contrast, all obvious phenomena are of the *kshara* world and its *prakriti* (nature) is *apara*. This nature pertains to *bhoomi, jalam, vayu* (the solid-liquid-gaseous phases of matter, *akasa* (space) and *agni* (radiation and its transmission from one place to another). It was discussed how the master medium can host solid, liquid or gas, and now it is also mentioned about radiation and space. Along with these factors, the mind, the intellect, and the subjective 'I' (*ahamkara* - ego), the base on which the mind-intellect-combine works; together form the body of a human being.

The Highest Entity expresses itself through *akshara* by primordial rhythm giving rise to creation, the resonance of which disperses through the entire universe. *Akshara* forms the *kshara* world through the fabric of space (*avyakta*) which, being a continuous substrate can only be in one of the three states – *vikasa* (expanding), *sankocha* (contracting), or *sama* (balanced). These represent the three *gunas*, the *rajas, tamas* and *sattva* respectively. Various forms of this incessant rhythm are on stage throughout the cosmos right from the microcosm to the macro level. The phenomena in the nuclear strata, chemical changes and every other development including the evolution of life are all born from these *gunas* (flavours or tendencies).

The subjective 'I' born out of the *apara* nature does not succeed in realizing, till the last phase of the evolution, the Supreme Entity which has both this *para-apara* nature. It is only after the formation of the frontal lobes of the brain that man becomes capable of such a realization, but being bound by his own temperaments, he hesitates to tread the way.

The universe remained in the embryo form (bīja rupa) in the *Purushottama* (highest entity or *Isvara*) and unfolded through the *akshara* out of which all the bodies in the universe are born. The most condensed state of the pulsation knows no time, space or motion. It is the pre-big bang period, and *akshara* itself is one with the *Purushottama* and there is only One entity. It is where no known physical law applies; it is called a 'discontinuity'. The will to expand is, at that instant, asleep in it. It wakes up to the start of expansion, marking the beginning of creation. The *Purushottama* and *Akshara* are perfectly described in the ancient *santi matra*:

Om purnamidam purnamadam purnal purnamudachade
Purnasya purnamadaya purnavevavasishyathe
Om santi, santi, santi hi..
(Om. That is Complete (Perfect), This is Complete (Perfect)
From That Completeness, This Completeness Arises (Awakes, Comes Forth)
From That Completeness, when This Completeness is 'Separated'
What remains is Completeness (Perfection)
Om. Peace, peace, peace be.)

Therefore the *Purushottama* is inferred to be the *ādikarana* (the first cause), *jeevakarana* (the cause of life), *ekam* (the unified) and *avyayam* (the eternal).

Unified force is just another name for the fundamental energy of the Supreme which becomes pluralized with the arising of *akshara.* The interpretation of this plurality is possible only if one accepts the presence and role of *akshara.* But it is seemingly not yet convinced in modern scientific disciplines; science continues to define space as vacuum and differentiating between space and matter, viewing them as unrelated entities and conceiving the universe only in terms of the matter in it. It is only beginning to recognize that there is a fabric of space, but has still not reached the stage for solving many a riddle including the one regarding the unification of fundamental forces.

The three gunas - sattva, rajas and tamas - arising from prakriti ties the forever dehi (purposeful life force, energy) to the body.

Bhagavad Gita 14:5

The word *'guna'* also means 'the rope that ties'. The meaning explains what role the *gunas* play in building up the universe. Oscillations exist in *avyakta* as deviations to either side of the neutral and maintain steady resonance with its surroundings. The situation can be viewed as the *gunas* binding the energy of the oscillation (matter wave) and preventing it from evening out with the surrounding *avyakta*. It's nature that enables it to pulsate and resonate is denoted by 'Om' (Aum); it provides stability to the wave form of all matter and radiation manifested in *akshara*, and also imparts stable cyclical (resonant) nature to the pulsation of the universe. *Rajas* signifies the tendency to expand, to be in excess, and to dominate. *Tamas* signifies the tendency to contract, to be lacking and to be recessive. (For example, visualize a simple harmonic pulsation and one can get a glimpse of two opposite phases plus an additional factor of stability.) *Sattva* signifies the tendency to be stable, to be at peace and to be neutral. The *gunas* 'tie' these oscillations together by making use of their deficiencies in resonance-fitness. This can be either merely quantitative (as in the case of the build-up of compounds, molecules, mountains, planets, stars and constellations) or qualitative (as in the case of biological assemblages).

It cannot definitely be made out how these three tendencies in oscillation tie the energy within the quark as its internal wave form, because the exact physical nature of the quark is unknown. It is too tiny and perceived as a 'point particle'. But in the next level, there is more direct and definite evidence of three opposite characteristics. Three quarks are required to form a neutron or a proton. And these quarks are required to be of three different 'colors' to form them. Physics does not know what these three colors are (they do not represent actual color even though they are named red, green and blue), but only that they represent three different and opposite characteristics. It is presently thought that these are three anti-symmetric combinations by which the nucleus is stabilized. It is also possible that these represent the quarks in three opposite nature of oscillation phases, brought together to form a stable nucleus. The quarks also 'madly' exchange these colors within the nucleus, which again indicate changing waveforms, components of which are shared between them to hold the nucleus of the atom as a single entity. However these suppositions are hypothetical and until physics finds some way to perceive the structure of the quark, the true physical nature of these wave forms will remain unknown.

In the next level the three tendencies repeat, but this time these are well known to physics. The constituents of the atom – the proton, electron and neutron – perfectly correlate with the three tendencies as positive, negative and neutral – two opposite phases plus the stability factor. They are the building blocks of the various elements.

In the fourth level also the three tendencies present themselves, again correlating perfectly with the three *gunas*. Inadequacy or excess (*tamas* or *rajas*) leads to bonding and formation of compounds, and elements show tendency to either grab electrons or provide them to other atoms. Some are in the relatively neutral (*sattva*) state. But it does not stop there, it continues; the same 'emotions' result in building up the entire spectrum of bodies, large and small in the universe.

The entirety of *akshara* takes part in the grand pulsing act of the universe. At no point in the universe therefore does the consistency of its fabric of space, *avyakta,* remain the same. From the big bang onwards the consistency of *avyakta* has steadily changed. As a result, an oscillation that enjoys a happy resonance slot goes out of tune in the course of time naturally, meaning the *guna* of its overall expression changes in character (color) or earnestness (brilliance) or both. So the energy or the purposeful force trapped as the oscillation makes use of the first chance it finds to team up with other oscillations that display a problem complementary to its own.

Prakriti (nature) gives birth to all and everything in the world under My authority; it is in the same way that change continues to go on in the world.

Bhagavad Gita 9:10

This is an important and simple explanatory verse. At the behest of the Supreme Force, the *prakriti* becomes splendorous. This splendour is manifest as interaction between the waves that form in the *avyakta* based on the dialectic nature of *akshara*. The Supreme Purusha is therefore, prima facie, the first cause of everything in the world. But *prakriti* (nature) is responsible for all the actual scripting and execution of the play. *Prakriti* creates the universe in an action reaction pattern. There are no miracles obviously disrupting the laws of *prakriti* (nature). Birth and death are inevitable; there is no wishing it all away. Light and darkness are the two sides of truth. If there is one, there is also the other. If *akshara* does not exist the world cannot.

Know that prakriti as well as purusha are beginning-less (forever). Also know that (all) emotions and gunas are born out of prakriti.

Bhagavad Gita 13:20

It is customary to perceive God as the Father and Nature as the Mother in many cultures. The cosmology propounded by the Upanishads has three levels to reality. The highest level is of the *Isvara (Purushottama)*. Next is *akshara* which is the combination of the dialectical fabric of space (*avyakta*) plus its resonance (represented by Om). The material world, the third and explicit level is a manifestation of the interactions of both. *Purushottama* provides the seeding for the *kshara* world at the start of the primary pulse which manifests as the universe. These two – *Purushottama* and *Akshara*, can also be described as *purusha* and *prakriti* respectively – the Universal Father and Mother. Both *purusha* and *prakriti* are beginning-less, meaning they outlive space and time. It is only the material world that cyclically comes into being and gets wiped out. *Prakriti* is dialectical because it has the three *gunas* – tendencies – two of which are mutually opposed, the third serving as transitional platform which is thus more or less free from the dialectical nature and correspond more with the characterless *Purushottama*. In the *avyakta* these *gunas* represent the expansion (*rajas*), contraction (*tamas*) and stable (*sattva*) phases, not only pertaining to the general pulsation of the universe but also to all oscillatory waves (both matter and radiation); in the *kshara* world their various permutation combinations are reflected as numerous qualities. Action-reaction patterns born out of these *gunas* are formed in *prakriti.*

Prakriti is said to be the reason for the cause-effect relationships within the body. Purusha is said to be the reason for all experiences.

Bhagavad Gita 13:21

The interaction between *purusha* and *prakriti* creates the universe. *Purusha* drives *prakriti* into cyclic expansion and contraction. The resultant variations in *prakriti* create and wipe out the material universe over and over again. The manifest universe can be viewed as a field for cause-and-effect interactions which has its origin from the background. The same action-reaction patterns work within the human body. *Purusha* resides in the body and illuminates all experiences resulting from these cause-effect interactions because it is the basic life force.

Prakriti involves both types of nature – the *para* and *apara*; both manifest and background nature. Events in the material world and our equipment to handle them are created by *prakriti*. The biosphere and the entire universe rest in *prakriti*, and everything are sustained by interactions at multiple levels by nature itself.

When one visualizes all apparently separate creations as residing in one, and (at the same time he sees) that same unity spread to fill the entirety, that person becomes one with brahma.

Bhagavad Gita 13:31

There is infinite diversity in creation. Even among the same genus and species, every individual is different. But all entities reside in *akshara* as waves in an ocean, after being produced by the ocean till dissolution in it.

The Divine One, the purusha without form, the One who is same both outside and inside, the One without birth, the One who is not prana or the mind, the splendorous; that One is definitely above even the noble akshara.

Mundaka Upanishad 2.1.2

So the logical conclusions about that Entity are: the Supreme One, often called God and denoted as the Supreme Purusha in the Vedas, is beyond even the substrate that forms the material world, and so beyond form or name. As It is everywhere it has no form, and no inside or outside. As there is no other, It has no name for identification. As It is eternal It is not born, nor does It ever die, It remains incredibly ancient and at the same time newest of new. It cannot be identified as individual life, nor can It be identified inside the mind. But It is the reason for the splendour of anything and everything in the universe. And It is the ultimate level, the Supreme Entity, greater than even the noble *akshara*.

This entire world is pervaded by Me through the template of avyakta; all beings exist in Me, but I do not (exclusively) dwell in them.

Bhagavad Gita 9:4

The Supreme Entity, though omnipresent, is indiscernible through the human faculties of senses. That Entity cannot be restricted into dwell in any of these beings, which also challenges the claim of any person who parades as God. It also rules out any God with a specific name or other human characteristics. 'Humanization' of God is not a Vedantic concept. Neither does it encourage idol worship.

These few verses shed light on how these sharp ancient thinkers perceived the creation and working of the universe. Their insights also point directly to the presence of a Supreme Entity that pervades everything in the universe and acts as life force for all creations. Why that entity is not evident to the sense organs is due to the in-between template of *akshara*. But without the latter invisible substrate the universe itself would not be possible.

Actually modern science is just one step behind this cosmological standpoint. It has been proved that matter and energy are one and the same; space is now considered as a field with action-reaction capability and not 'empty'; that the universe began from the 'explosive' start of a pulsation has been accepted. There are numerous evidences for this unobservable medium (*akshara*) too. The trinity formed by the three levels of reality – *kshara, akshara* and *Isvara* – can serve as the all-encompassing model for the universe.

******

The few verses discussed so far in this book provide only a brief outline of Vedantic cosmology, to correlate with what Physics has inferred so far. Readers interested in Vedanta can find more verses and read a much detailed study in the author's research and commentary on the Bhagavad Gita:

Bhagavad Gita: Modern Reading and Scientific Study

By

C. Radhakrishnan

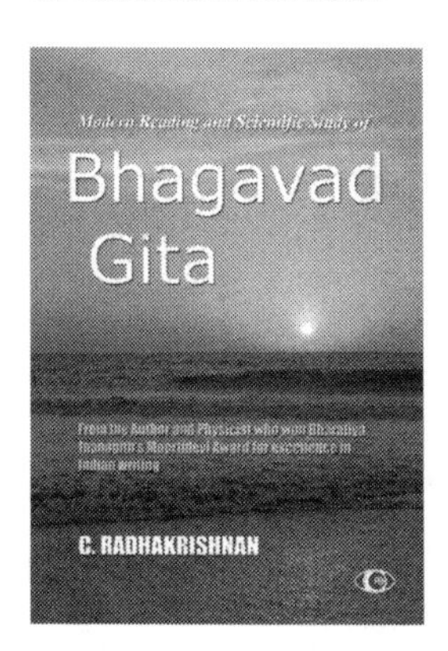

The book presents in-depth study of ancient Indian philosophy, bringing out the essence of Vedanta in its clear and pristine form devoid of any religious consideration or superstition.

The emotional state of the observer, experimenter or applicator is no concern of modern science. Parā vidya, on the other hand, in addition to seeking knowledge, presents the theory and practice of internal fulfillment. In fact, the knowledge it offers cannot be imbibed unless one's mind is under control. Relying on superstition is not the answer. Rigid religious texts may not answer complex questions unless one learns for oneself how to work out answers and live accordingly. So the right combination of parā vidya (true knowledge of oneself) and modern science can be considered a viable antidote to the ills of the world.

This mega work establishes how the wisdom provided in the Gita can help anyone lead a successful and happy life in today's modern world irrespective of religion or nationality, successfully integrating Vedantic ideas with contemporary life in unbiased, simple and practical manner. (No self-help books required anymore!).

The book contains Sanskrit transliteration of the Gita into lossless roman script with English translation of every verse, plus detailed commentary and discussion. Sanskrit terms used in the Gita are retained, and all concepts however deep are explained lucidly making it very easy for objective interpretation even for the beginner.

Related work from the Author available from Amazon:

Deep Within

Science fiction novel based on Indian philosophy.

A young couple on their scientific research is about to make a huge breakthrough – a revolutionary technology that could benefit humankind as never before and herald a new world.

But every gift from science can be misused...

A global criminal gang is after them, trying

to hold them at ransom, ready to do anything to grab that technology for the final war of dominance over the world.

But the couple fight back with their mind and heart and power derived through an ancient philosophy.

The mind is the ultimate weapon....

Famous Literary Fiction works of the Author available from Amazon:

Birds That Fly Ahead

This book is the English translation of the Malayalam work *Munpeparakkunna Pakshikal*, a bestseller running into more than a dozen editions and winning as many as seven major Awards.

Presenting in its background many an untold part of an extremist upsurge that took place in India four decades back and the way it has influenced the fabric of life ever since, the novel deeply explores human nature and thought in the drive for a happier and saner world.

After this earth and the human community came into being, so many great souls have flown ahead, showing and leaving open the routes to worlds of lasting happiness. But humankind, still, either unknowingly or wilfully, prefers the path to the butcher-houses we ourselves built to reach there!

It is a modern Indian classic, the work of a master craftsman, different in style, rich in content, revealing and invigorating.

Heart-Rending Times

How much can a woman take without breaking down? A great lot more than ever imagined is what this book portrays. Most epics and, later, classical works of literature have women made to walk on fire, but Anuradha lives in a veritable inferno. At critical moments of despair and pain she is surprised by the array of strange personalities emerging from deep within her own self.

In ways how unusual can a man love a woman? How selfish and cruel can intense love turn when spurned? What does vengeance do to the person in his teens harbouring it? And what all drives a mother to act in ways so bewildering in circumstances never faced before in human history?

Now For a Tearful Smile

This is a unique and poignant love story of two very special people living under assumed identities for different reasons; and a thriller chasing the intriguing tentacles of espionage activities and the untold stories behind political assassinations in India.

This landmark novel, a socio-political silhouette of contemporary India, is a poignant juxtaposition of materialism and spirituality, contemporary geopolitics and history in the making, India and the world, urban and tribal cultures, tragedy and comedy, saints and sinners, displacement and assimilation, cyanide tablets and self-healing, arms and man, war and peace, hospitals and hermitages... innumerable are the strands that form the warp and weft of this literary milestone. It thrills, threatens, moves, shocks, calms and exhilarates!

References (Physics)

Walter Greiner (2001). Quantum Mechanics: An Introduction. Springer. ISBN 3-540-67458-6.
Brian Greene, The Elegant Universe, page 104 "all matter has a wave-like character"
Davisson, C. J.; Germer, L. H. (1928-04-01). "Reflection of Electrons by a Crystal of Nickel". Proceedings of the National Academy of Sciences of the United States of America 14 (4): 317–322. ISSN 0027-8424. PMC 1085484. PMID 16587341.
M. Born & E. Wolf, Principles of Optics, 1999, Cambridge University Press, Cambridge
H. Merimeche (2006). "Atomic beam focusing with a curved magnetic mirror". Journal of Physics B 39 (18): 3723–3731. Bibcode: 2006JPhB... 39.3723M.Doi:10.1088/0953-4075/39/18/002.
Razavy, Mohsen (2003). Quantum Theory of Tunneling. World Scientific. pp. 4, 462. ISBN 9812564888. Serway; Vuille (2008). College Physics ***2*** *(Eighth ed.). Belmont: Brooks/Cole. ISBN 978-0-495-55475-2.*
Bjorken and Drell, "Relativistic Quantum Mechanics", page 2. Mcgraw-Hill College, 1965.
Barger, Vernon; Marfatia, Danny; Whisnant, Kerry Lewis (2012). The Physics of Neutrinos. Princeton University Press. ISBN 0-691-12853-7.
Ahmad, Q. R.; et al. (SNO Collaboration) (2001). "Measurement of the Rate of ve + d → p + p + e− Interactions Produced by 8B Solar Neutrinos at the Sudbury Neutrino Observatory". Physical Review Letters 87 (7). arXiv:nucl-ex/0106015.
Richard P. Feynman (1970). The Feynman Lectures on Physics Vol I. Addison Wesley Longman.
David Nikolaevich Klyshko (1988). Photons and nonlinear optics. Taylor & Francis. p. 126. ISBN 2-88124-669-9.
Milton K. Munitz (1990). Cosmic Understanding: Philosophy and Science of the Universe. Princeton University Press. p. 132. ISBN 0-691-02059-0. The spontaneous, temporary emergence of particles from vacuum is called a "vacuum fluctuation".
Astrid Lambrecht (Hartmut Figger, Dieter Meschede, Claus Zimmermann Eds.) (2002). Observing mechanical dissipation in the quantum vacuum: an experimental challenge; in Laser physics at the limits. Berlin/New York: Springer. p. 197. ISBN 3-540-42418-0.
Christopher Ray (1991). Time, space and philosophy. London/New York: Routledge. Chapter 10, p. 205. ISBN 0-415-03221-0.
Astrid Lambrecht (Hartmut Figger, Dieter Meschede, Claus Zimmermann Eds.)

(2002). Observing mechanical dissipation in the quantum vacuum: an experimental challenge; in Laser physics at the limits.
Modern Electrodynamics (1st ed.). Cambridge University Press. 2012-12-24. pp. 46–48. ISBN 9780521896979.
Peskin, M.E., Schroeder, D.V. (1995). An Introduction to Quantum Field Theory, Westview Press, ISBN 0-201-50397-2, p. 80.
Mandl, F., Shaw, G. (1984/2002). Quantum Field Theory, John Wiley & Sons, Chichester UK, revised edition, ISBN 0-471-94186-7, pp. 56, 176.
Bayfield, James E. (1999). Quantum evolution: an introduction to time-dependent quantum mechanics. New York: John Wiley. p. 62. ISBN 9780471181743.
Hiroyuki Yokoyama & Ujihara K (1995). Spontaneous emission and laser oscillation in microcavities. Boca Raton: CRC Press. p. 6. ISBN 0-8493-3786-0.
Marian O Scully & M. Suhail Zubairy (1997). Quantum optics. Cambridge UK: Cambridge University Press. p. 1.5.2 pp. 22–23. ISBN 0-521-43595-1.
Timothy Paul Smith (2003). Hidden Worlds: Hunting for Quarks in Ordinary Matter. Princeton University Press. ISBN 0-691-05773-7.
Cho, Adiran (2 April 2010). "Mass of the Common Quark Finally Nailed Down". http://news.sciencemag.org. American Association for the Advancement of Science.
Watson, A. (2004). The Quantum Quark. Cambridge University Press. pp. 285–286. ISBN 0-521-82907-0.
Reynolds, Mark (Apr 2009). "Calculating the Mass of a Proton". CNRS international magazine (CNRS) (13). ISSN 2270-5317.
Weise, W.; Green, A.M. (1984). Quarks and Nuclei. World Scientific. pp.65–66. ISBN 9971-966-61-1.
Ball, Philip (Nov 20, 2008). "Nuclear masses calculated from scratch". Nature. Doi:10.1038/ news.2008.1246.
Particle Data Group. "2010 Review of Particle Physics.
Strassler, M. (12 October 2012). "The Higgs FAQ 2.0". ProfMattStrassler.com. [Q] Why do particle physicists care so much about the Higgs particle? [A] Well, actually, they don't. What they really care about is the Higgs field, because it is so important. [emphasis in original]
R. Nave. "The Xi Baryon". HyperPhysics.
Sean Carroll, Ph.D., Cal Tech, 2007, The Teaching Company, Dark Matter, Dark Energy: The Dark Side of the Universe, Guidebook Part 2 page 46, Accessed Oct. 7, 2013, "...dark matter: An invisible, essentially collisionless component of matter that makes up about 25 percent of the energy density of the universe... it's a different kind of particle... something not yet observed in the laboratory..."
Freeman, K.; McNamara, G. (2006). In Search of Dark Matter. Birkhäuser. p. 37.

ISBN 0-387-27616-5.
Babcock, H, 1939, "The rotation of the Andromeda Nebula", Lick Observatory bulletin; no. 498
Collins, G. W. (1978). "The Virial Theorem in Stellar Astrophysics". Pachart Press.
Refregier, A. (2003). "Weak gravitational lensing by large-scale structure". Annual Review of Astronomy and Astrophysics 41 (1): 645–668. arXiv:astro-ph/0307212.
Masters, Karen (September 2002), What is the Origin of Spiral Structure in Galaxies
Michelson, Albert A.; Morley, Edward W. (1887). "On the Relative Motion of the Earth and the Luminiferous Ether". American Journal of Science 34: 333–345. doi:10.2475/ajs.s3-34.203.333.
Albert Einstein (1905) "Zur Elektrodynamik bewegter Körper", Annalen der Physik 17: 891; English translation On the Electrodynamics of Moving Bodies by George Barker Jeffery and Wilfrid Perrett (1923); Another English translation On the Electrodynamics of Moving Bodies by Megh Nad Saha (1920).
Forshaw, Jeffrey; Smith, Gavin (2014). Dynamics and Relativity. John Wiley & Sons. ISBN 978-1-118-93329-9.
FitzGerald, George Francis (1889), "The Ether and the Earth's Atmosphere", Science 13 (328): 390, PMID 17819387
Lorentz, Hendrik Antoon (1892), "The Relative Motion of the Earth and the Aether", Zittingsverlag Akad. V. Wet. 1: 74–79
1.Feynman, Richard (1985). QED: The Strange Theory of Light and Matter. Princeton University Press. ISBN 978-0-691-12575-6.

Made in United States
Troutdale, OR
03/24/2024